Learn to Solder

Brian Jepson, Tyler Moskowite, and Gregory Hayes

MAKER MEDIA™
SAN FRANCISCO, CA

Learn to Solder

by Brian Jepson, Tyler Moskowite,, and Gregory Hayes

Editors: Brian Jepson and Shawn Wallace   **Interior Designer:** David Futato
Cover Designer: Brian Jepson   **Photographer:** Gregory Hayes

May 2012:   First Edition

Revision History for the First Edition

April 27, 2012:   First Release
November 30, 2012:   Second Release
August 5, 2015:   Third Release

See *http://oreilly.com/catalog/errata.csp?isbn=9781449337247* for release details.

978-1-449-33724-7

[LSI]

Contents

Welcome

Welcome to Make:'s *Learn to Solder*, the companion book to our Getting Started with Soldering kit available at Maker Shed (*http://www.makershed.com/*). Congratulations on taking a big step into the world of DIY electronics. Once you get the hang of soldering, you can put together some of the many great kits that are available, fix electronics that are broken, and build inventions of your own. With this kit, you'll learn how to:

Prepare and clean your soldering iron

Assemble electronic circuits from kits

Solder reliable connections that are neat and clean

Correct soldering mistakes you've made

If you don't have our soldering kit, you can purchase the tools individually from Maker Shed (*http://www.makershed.com/*).

Basic Tools

There's a lot of great stuff in the box, and before you start using it, here's a tour of what you'll find in there. Figure P-1 shows the soldering tools you'll be using most of the time. Note that the parts shown here may look slightly different than what's in the box (but they will certainly work identically).

Deluxe soldering iron
(Top left) This soldering station includes a variable temperature controller, a cleaning sponge, soldering iron, and a ringed holder.

Crosscut pliers
> (Left) You'll use these to trim away excess leads after you solder components in place.

Tube of lead-free solder
> (Bottom) This is enough solder to get you started and keep you busy over many projects. See "Lead-Free vs. Leaded Solder" on page 7 for some information on lead-free solder.

Helping hands
> (Right) For those times when you need a third or fourth hand, the helping hands let you hold items steady while you solder.

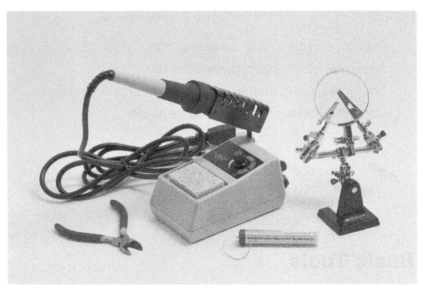

Figure P-1. *Basic soldering tools*

Lead-Free vs. Leaded Solder

Solder is a metal alloy that you fuse to your electronic connections using lots of heat (which the soldering iron delivers). For many years, much of the solder in use was made with a combination of lead and tin. In the EU, the use of lead in solder is now prohibited in most cases, due to the toxicity of lead. Lead-free solder has a slightly higher melting point than lead-based solder.

Among makers and engineers, you'll find a lot of different opinions regarding lead-free solder: many consider it slightly harder to work with, and some worry that electronics built with lead-free solder may develop problems over time. We believe that lead-free solder is the best choice for simple electronics projects, but if you're building something (like a spaceship or pacemaker), you'll no doubt want to do a lot of research into which solder is right for you.

Advanced Tools

Most of the time, you'll only need the basic tools to get things done. But when you need to replace your soldering iron tip, correct a mistake you made while soldering, or need a tool to help keep components from overheating, you'll need the items shown in Figure P-2.

Desoldering wick
    (Bottom left) Use this to wick away excess molten solder.

Desolder pump
    (Top left) This pump will suck up molten solder when you have a lot of solder to remove.

Soldering tools
    (Center right) The scrapers, brush, and slotted probe come in handy when you need to move solder around or precisely position a component.

Extra tips

(Right) Tips don't last forever, and they come in different shapes. When you've worn out your tip, or if you need a conical/pointed tip instead of a chisel tip, use one of these. A chisel tip is best for the type of soldering you'll learn in this book. Because it's wider than a pointed tip, it carries more heat to your solder joint, making it easier to solder. A pointed/conical tip, or at least a slightly more narrow chisel tip, is good for working with components that can be easily damaged by heat, or when soldering surface-mount components, which have very small leads.

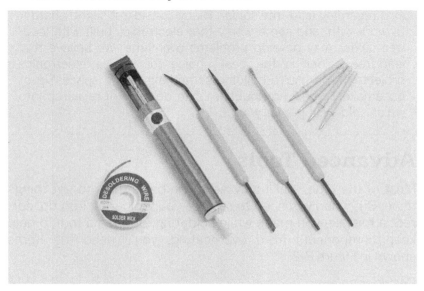

Figure P-2. *Advanced soldering tools*

Project 1: Learn to Solder Badge

Our Learn to Solder Badge Kit (shown in Figure P-3) has been used to teach thousands of people of all ages how to solder at Maker Faires across the country. It's a simple, fun way to learn how to solder and also to teach others. After you build the ones included in the box, you can order more from *http://make-rshed.com* and teach all your friends to solder.

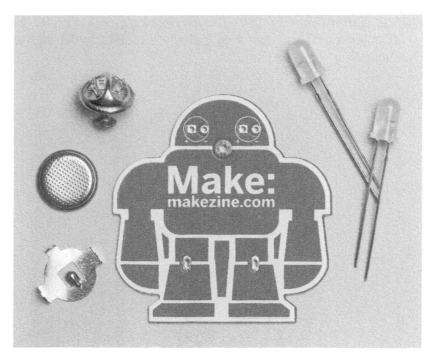

Figure P-3. *Learn to Solder Badge, 2012 model*

Pin and clutch
   (Top left) This is what holds the pin to your clothing.

Blinking LED
   (Top right) This is an LED (Light-Emitting Diode) with a twist. Normal LEDs keep shining as long as you give them power. Like other LEDs, this LED has one element (red), but unlike other LEDs, that element is under the control of a small integrated circuit (IC) embedded within the LED. The IC causes the element to blink.

Printed Circuit Board (PCB)
   (Center) The 2012 Learn to Solder Badge features a friendly robot.

Battery holder
   (Bottom left) This keeps the battery on the PCB.

CR1220 battery
 (Center left) This "coin cell" battery supplies power to the pin.

1/Getting the Workspace Ready

Before you can start soldering, you need to get your iron ready. This involves a few steps. First, you'll have to attach the tip to the soldering iron. Then, you'll need to prepare the station: make sure your sponge is moistened, set the temperature, and "tin" the tip by applying some solder to it.

At the end of this chapter, your soldering iron will be assembled, clean, and hot enough to begin soldering.

Solder contains harmful chemicals. Wash your hands after soldering as well as after handling solder or the solder station sponge. Though it may be tempting to snack while you're working, keep any food or drink away from your soldering area.

1: Attach the Tip

Locate the soldering iron (make sure it's unplugged and cool), a soldering iron tip, and the tip nut that holds the tip in place. Insert the tip into the soldering iron as shown in Figure 1-1, and push it in as far as it will go. It will not require much force to insert it.

 Make sure the soldering iron is unplugged and cool.

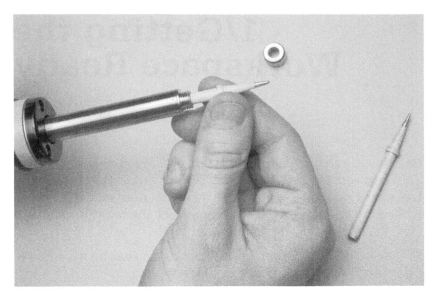

Figure 1-1. *Inserting the tip into the tube*

2: Secure the Tip

Slide the tip nut over the tip, and screw it in place. You can use pliers to tighten it (see Figure 1-2). Be careful not to overtighten it, since you will eventually need to replace it when the tip gets worn from all the projects you'll be making with it.

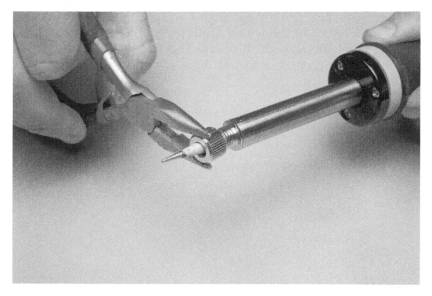

Figure 1-2. *Making sure the tip stays put*

Over time, the tip nut might become loose. Let the iron cool down before you tighten it.

3: Wet the Soldering Station Sponge

You'll need to perform this step (and the remaining ones in this chapter) at the beginning of each soldering session.

As you use the soldering iron, you're going to be wiping the tip a lot, so you need to keep the solder station sponge moist. Pour a little clean, fresh water on the sponge as shown in Figure 1-3. Don't soak it, but make sure it's completely moistened.

If your water has a high mineral content, use distilled water for wetting the sponge.

Make sure the solder station is unplugged when you wet the sponge this way. The sponge is removable, so you could also wet it away from the solder station.

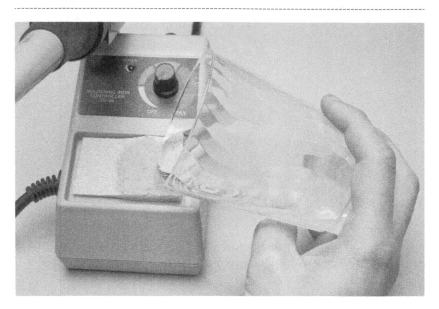

Figure 1-3. *Wetting the sponge*

4: Set the Station's Temperature

Plug your soldering iron in, and set the heat setting to the mark shown in Figure 1-4 (pointing at the 3 o'clock position on the dial). Give the iron a couple minutes to warm up.

Figure 1-4. *Setting the temperature*

Some tips will give off a little smoke when you heat them up for the first time. This is typically due to the burnoff of residual oil that the manufacturer applied to resist rusting.

The iron will get very hot, and certainly hot enough to burn you. Don't touch the tip and don't touch the tip to anything other than solder or components you are soldering together.

5: Wipe the Iron's Tip

Wipe the tip on the sponge as shown in Figure 1-5, being sure to turn the iron a few times as you wipe it, so you wipe all of its surfaces. It will make a sizzling sound as the water in the sponge comes into contact with the tip. The figure shows a sponge that's been used a few times. Notice the little blobs of solder that have come off the tip over time. A clean tip is critical to successful soldering. Even a small layer of oxidized material or other crud will limit the amount of heat that's transferred to the component you're soldering. Wipe the tip often, and keep the tip clean.

Figure 1-5. *Wiping the tip clean*

6: Tin the Iron's Tip

Before you begin soldering components, you should put a thin layer of solder on the tip as shown in Figure 1-6. This will have two effects: burning off any crud that shouldn't be there, but also providing a thin layer of liquid solder. This layer makes for a very effective heat transfer between the soldering iron and whatever you're soldering. Give the tip one more quick wipe on the iron and place the soldering iron into the holder until you need it.

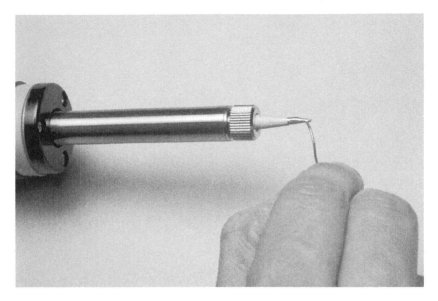

Figure 1-6. *Tinning the soldering iron tip*

Depending on the composition of the solder you use, a small puff of smoke may appear as the rosin flux in the center of the solder is activated. The rosin serves two purposes: to clean the soldering iron (and the joint that you are soldering) and to help the solder flow freely. Avoid inhaling this smoke.

2/How to Solder

With your workstation all set up, you're ready to start soldering! Your first project is the Learn to Solder Badge ("Project 1: Learn to Solder Badge" on page 8). This chapter spells out the tips and techniques for soldering effectively, safely, and neatly.

Working with Solder

Solder is made of metal with a core that contains flux, which cleans the connection as you solder. When the metal melts, the flux begins to flow onto the joint. Some solder lead-based, but the solder in this kit is lead-free. To melt solder, heat it with the soldering iron as shown in Figure 2-1.

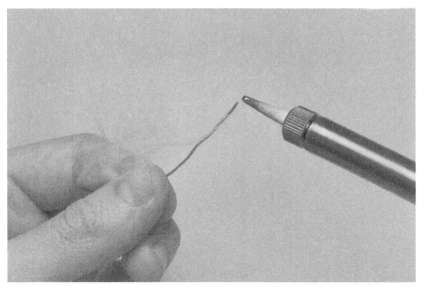

Figure 2-1. *Melting some solder*

Try melting some solder, but be careful not to drip it on yourself or anything other than a work surface. You can try pushing tiny balls of solder around on your work surface to see how it flows.

If you have some bare hookup wire or similar metal, try heating it up by holding the soldering iron to it (don't hold the wire while you're heating it). Touch a piece of solder to the wire, but choose a point that's an inch or so from where the soldering iron is touching. How close do you have to get to the soldering iron tip before it melts? (For some background on the composition of solder, see "Lead-Free vs. Leaded Solder" on page 7.)

Keeping the Circuit Board from Moving

When you're soldering, you've got a lot to juggle: the soldering iron, the solder, and the two things you are connecting to each other. The helping hands let you place the items you're soldering in a stable position. Use the clips to hold the item in place as shown in Figure 2-2. But once you start inserting items into the Learn to Solder Badge printed circuit board (PCB), you need some way to keep them from moving. "Placing a Component in the Board" on page 23 shows you how to do that.

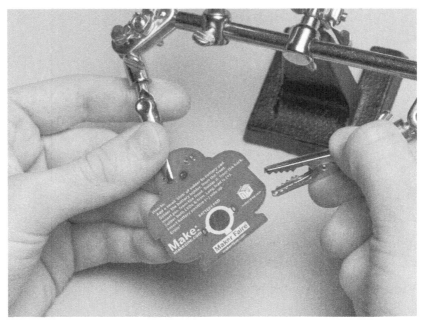

Figure 2-2. *Placing your Learn to Solder Badge PCB in the helping hands*

Tinning Solder Pads

You'll sometimes come across things that need a layer of solder to work right. For example, on the Learn to Solder Badge, you need to put a bump of solder in place to give the battery a snug fit. With solder pads that are this big, you need to heat the pad with the iron really well; the pad is so large that it's going to take longer to heat. It's best if you melt the solder by pushing it onto the pad rather than pushing it directly against the iron. This makes sure that the solder flows thoroughly over the pad. If the solder doesn't melt, try tinning the tip again first.

Start with a little blob as shown in Figure 2-3, and spread it around evenly. When you've got the desired thickness, first pull the solder away, then pull the soldering iron away. Doing it in this order will avoid leaving chunks of solder behind.

Figure 2-3. *Laying down some solder*

Knowing Which Way a Component Goes In

Before you place any components in the board, there's something you need to know. Some components don't care which way you put them in: resistors, some capacitors (such as the one that comes with this kit), and many other components fall into that category. On the other hand, LEDs are polarized; electrical current will only flow through them in one direction. So if you put your LED in backwards, it won't light up at all. You can determine an LED's polarity in a couple of ways: First, the longer of the two leads is the positive (+), and the shorter is negative (-). You can see the shorter lead on the right side in Figure 2-4.

Second, look closely at the bottom of the LED and examine the ring that bulges out around it. There is a flattened part of that ring that indicates the side of the LED that's negative (-).

With ICs, this also matters. Every pin has a specific function, and there will be one or more pins for positive (+) and one or more pins for negative (-).

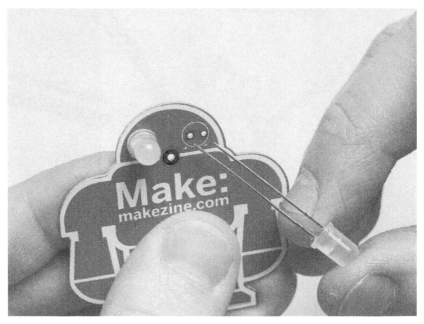

Figure 2-4. *Polarity—LEDs have it*

Placing a Component in the Board

You don't want components falling out of the board while you're soldering. In fact, you'll often insert a component with the board upside down (see Figure 2-5), then flip the board over when you put it in the helping hands. So the components have plenty of opportunity to fall out. To keep the component in place, bend the leads out as shown in Figure 2-6.

In some cases, you may need to quickly tack a component in place to keep it from moving. See "Stabilizing and Straightening Components" on page 24.

Don't try to place every component at once. Start with low-profile (shorter) components, solder them in one at a time, and move on to higher-profile components.

Figure 2-5. *Inserting an LED*

Figure 2-6. *Bending out the leads*

Stabilizing and Straightening Components

Although you won't be soldering an Integrated Circuit (IC) in the Learn to Solder Badge, you should know that ICs present a unique problem. They don't have long enough leads for you to bend out effectively, which means they are more likely to slip and slide while you're soldering them. So do the best you can with that (see Figure 2-7, which shows a 555 timer IC being soldered to a breadboard PCB), but flip the board back over after you solder the first joint. In most cases, you'll find that the IC has shifted. But since you only have one joint soldered, you can easily move the IC into the position you want. Once you've done this, flip the board over and solder the remaining connections.

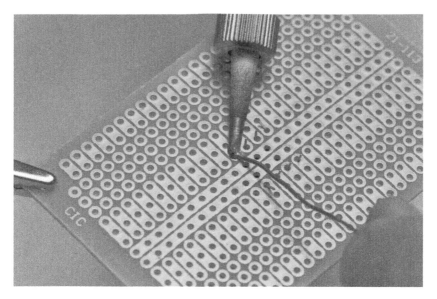

Figure 2-7. *Tacking down an Integrated Circuit*

Soldering a Component in Place

Wipe the soldering iron on the sponge and hold it so it's touching both the solder pad and the lead that protrudes through. Avoid using the soldering iron to melt the solder. Instead, heat up both the lead and the pad so that the solder melts when you touch it to them.

This must all be done quickly: within a second of touching the iron to the joint, push a small amount of solder into the joint, and let it flow around the joint as shown. Give it a second to flow, take the solder away, then wait a second, and take the iron away. With practice, you should be able to do all of this in three to four seconds per joint. Figure 2-8 shows how you'd solder the pin to the badge. Figure 2-9 shows an LED being soldered.

If you hold the iron to the component too long, you run the risk of damaging either the component or the solder trace on the PCB.

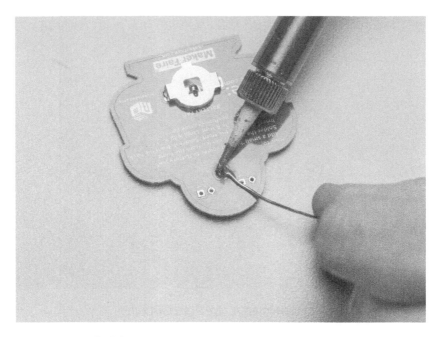

Figure 2-8. *Soldering a pin*

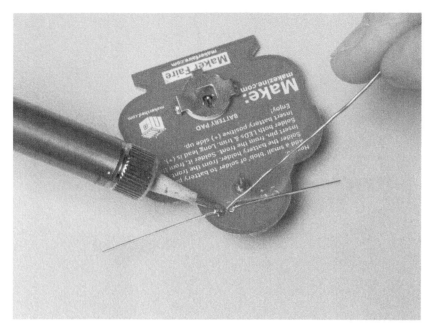

Figure 2-9. *Soldering an LED*

Trimming Your Leads

After you solder a component in place, you need to tidy things up. Grab the cutters from your kit, and trim the leads as close to the solder joint as possible (see Figure 2-10). Don't cut into the solder joint; the idea is to trim the excess wire lead.

When you clip it, the wire lead will fly away from the clippers very fast, and could injure someone. The best way to keep this from happening is to use one or two fingers to hold the lead you're clipping. With one finger, you can put enough pressure on it to keep it from flying away.

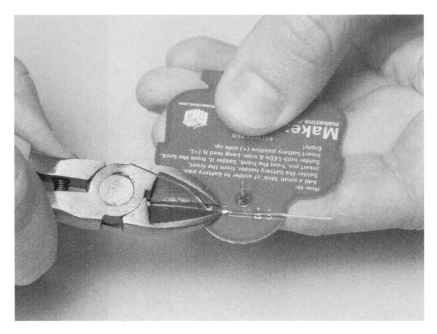

Figure 2-10. *Trimming leads on the Learn to Solder Badge*

Getting the Perfect Solder Joint

It's not hard to get a great solder joint, but it's also easy to be sloppy. Here are some pictures of good and bad joints to help you out.

Figure 2-11
Here's a good solder joint on the positive (+) pad: solder surrounds the joint and makes a little peak. This joint had just the right amount of solder and just the right amount of heat.

Figure 2-12
The joint on the negative (-) pad is sloppy. The solder didn't flow around the joint, and some of it is spread out across the lead. You can try to fix it by holding the iron to the joint for a couple seconds; the solder should flow down the lead and onto the joint.

Figure 2-13
The joint on the positive (+) pad has too much solder, and it's balled up. You can use the desoldering wick or solder

sucker to remove the excess. After you do that, hold the iron to the joint for a couple seconds to get the solder to flow around it.

Figure 2-14
Here are two joints for comparison: one blobby, one good.

Figure 2-11. *A good solder joint*

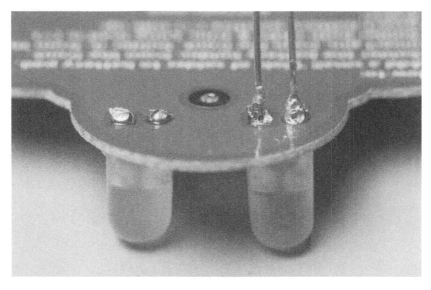

Figure 2-12. *A messy solder joint*

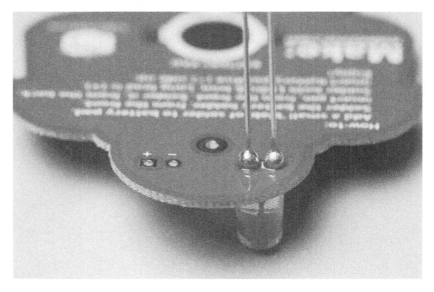

Figure 2-13. *A blobby solder joint*

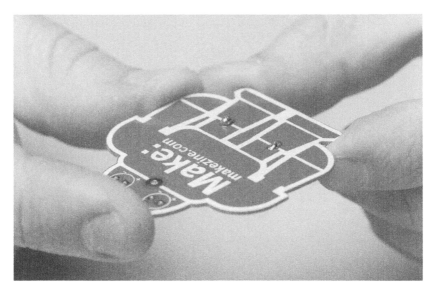

Figure 2-14. *Comparing two joints*

3/Desoldering

If everything went right the first time, you'd never learn anything. This section shows you how to recover from little soldering mistakes, and also includes a gallery of various solder joints so you can compare your work.

Desoldering is the "undo" command for soldering. With this technique, you can remove solder from a joint, allowing you to free the component so you can reorient it as needed. There are two tools for this: the desoldering wick (sometimes called desoldering braid) and the solder sucker.

The Desoldering Wick

Pull a small length of wick out of its spool (Figure 3-1). You don't need much, but if you don't use it all, you can wind the wick back into its spool.

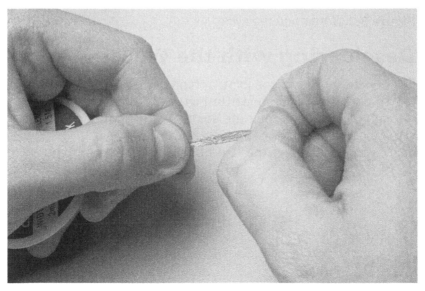

Figure 3-1. *Pulling out some wick*

Preparing the Wick

Before you use the wick, hold it as shown in Figure 3-2, and push inward. This will cause the wick to spread out, giving you more surface area to work with.

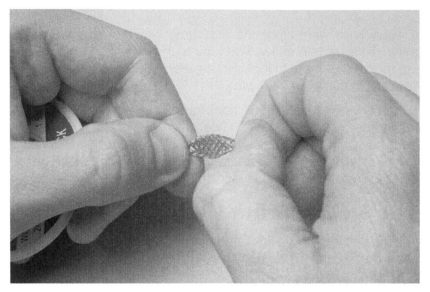

Figure 3-2. *Spreading the wick*

Desoldering with the Wick

Position the wick over the joint, and press down on the wick with the soldering iron as shown in Figure 3-3. The heat will pass through the wick and melt the solder. As this happens, the wick will soak up the solder, removing it from the joint. When you're done, you can trim off the used portion of wick with your wire snips.

Figure 3-3. *Desoldering*

 The wick will get very hot, so take your hands off it as soon as you press the soldering iron to it (the soldering iron will hold it in place).

Desoldering with the Solder Sucker

The solder sucker is another tool for removing solder. It's best for removing excess solder, and you can use it to clean things up a bit before you use the wick. To use the solder sucker, press the plunger down until it locks. Use the soldering iron to melt the solder around the joint, and quickly bring the solder sucker's nozzle to the joint (Figure 3-4). Press the button on the sucker, and it will pull the solder right up into the tube. When you press the plunger again, any solder in the tube will be ejected.

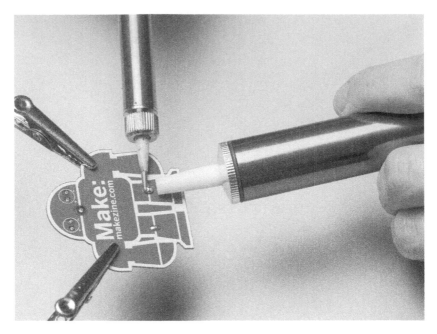

Figure 3-4. *Using the solder sucker*

4/What's Next?

Now that you've learned the ins and outs of soldering, it's time to do more. You can share this book with a friend so they can learn, or you can teach them yourself. And the world is full of wonderful things you can make now that you know how to solder.

Teach the World to Solder

Before you can teach someone to solder, you need to figure out what you want them to try making. You can purchase more Learn to Solder Badge Kits from *http://makershed.com*. Here are a few tips for teaching soldering to others:

Keep Your Hands Off the Tools
You're there to teach, not to solder. And your students will learn best if they always have their hands on the tools. If you grab the tools from them and do part of the work, they won't learn from it.

Be Patient
Try to remember what you went through the first time you tried to solder. It takes time, and you did things incorrectly, and maybe you even made mistakes that required desoldering. Let your students learn from their mistakes, and let them take their time.

No Food or Beverages
Even with lead-free solder, there are chemicals in the flux that aren't good for you. If you allow food or drink near the work area, it could get contaminated. Keep food and drink away from where you're working.

Make Sure There's Plenty of Light
The components you work with while you solder are small, and hard to see. Make sure to use a well-lit workspace for teaching.

Keep Spare Parts Around

Beginners will break things. Pick up some spare resistors, LEDs, and other common components, and have them on hand in case you need to replace a damaged component.

Make More Things

Maker Shed (*http://makershed.com*) carries many electronic kits that you can apply your newly-learned soldering skills to.

A/Going From Solderless Breadboard to PCB

If you want to design your own electronic projects, you'll generally use a solderless breadboard to prototype it. But what do you do when you're ready to make a permanent version of your project? This is where your newly-learned soldering skills come in handy. You can migrate your project directly to a *protoboard*, a special PCB for making prototypes into permanent creations.

The beauty of protoboard is that it mirrors the layout of a solderless breadboard, which you almost certainly will use for prototyping your electronic circuits. Once you have your project laid out and working on a solderless breadboard, you can easily transfer it to the protoboard. As with a solderless breadboard, the protoboard has rows and columns that are tied together. The layout is slightly different, though. Where the breadboard has two rails (one for positive, one for negative) on both the top and bottom as shown in Figure A-1, a typical protoboard its rails in the center of the board. You can find many variations on protoboards, including the Adafruit Perma-Proto boards we sell on Maker Shed (*http://www.makershed.com/pages/search-results?q=perma-proto&p=1*).

The circuit shown in this section is a simple blinking circuit using the 555 timer, the most popular IC ever manufactured. For a complete tutorial on electronics, and many experiments involving the 555 timer, check out Make: Electronics, 2nd Edition (Make:) by Charles Platt. For instructions on building this 555 Timer Blinky, see *http://makezine.com/projects/555-timer-blinky/*.

Figure A-2 shows the project from Figure A-1, but laid out on a protoboard instead of the solderless breadboard.

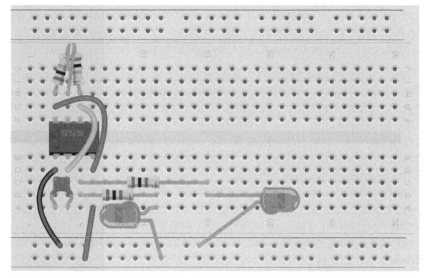

Made with **Fritzing.org**

Figure A-1. *The 555 blinky project on a breadboard*

The breadboard diagram in Figure A-1 was made with Fritzing, an open-source initiative to support designers, artists, researchers and hobbyists to work creatively with interactive electronics. For more information, see *http://fritzing.org/*.

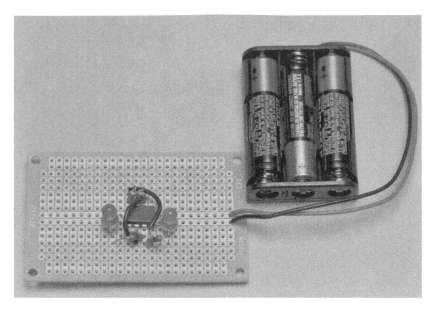

Figure A-2. *The 555 blinky project on a protoboard*

Soldering Jumper Wire

Jumper wire can be treated a lot like other components (see "Placing a Component in the Board" on page 23): make sure there's enough bare wire poking through so you can bend it outward enough to hold it in place. The notched probe included with this kit can be helpful with placing jumper wire in the right place (see Figure A-3).

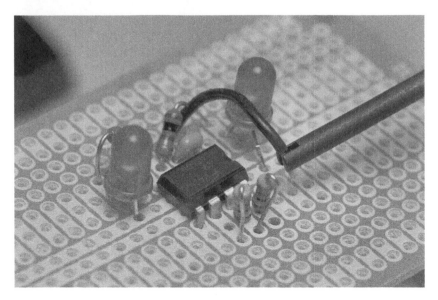

Figure A-3. *Placing the jumper wire*

Bridging Joints with Solder

When you're working with the protoboard, you will sometimes need to bridge joints that are close to each other. This is preferable to using jumper wire since your board remains uncluttered on the top. It doesn't always look great on the bottom, though, because you end up with a lot of solder in some places. Figure A-4 shows the start of a bridge: feeding the solder into the gap between solder joints. Figure A-5 shows a bridge being formed.

Figure A-4. *Feeding solder into a gap*

Figure A-5. *Forming a bridge across the gap*

Relieving Strain on Cables

If you have something heavy dangling off your board, such as a battery box, it's very likely that the solder joints will come undone. If you pass the wires through the large holes on the board before soldering them down (Figure A-6), you can add some strain relief (Figure A-7) that will help prevent this from happening. If you have a small drill, you can also drill out a couple of holes close to where you are going to solder the connection, and loop the wire through there. This will be even more rugged than using the larger holes.

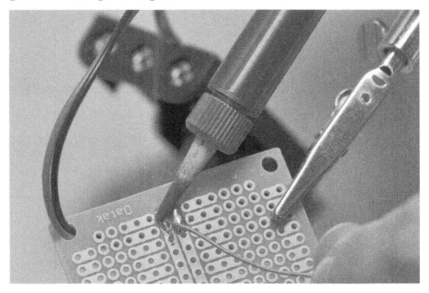

Figure A-6. *Soldering the battery holder wires*